I0393785

Project MIND
<u>M</u>ath <u>I</u>s <u>N</u>ot <u>D</u>ifficult

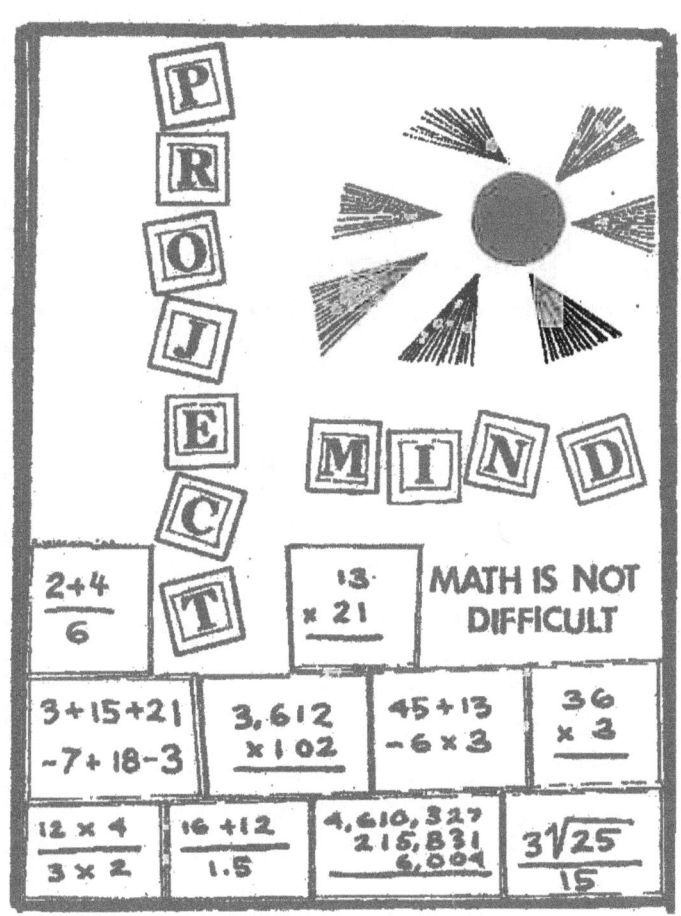

Fourth Grade
Mental Math Flash Cards
Hui Fang Huang "Angie" Su, Ed.D.
Project MIND, Inc.

Copyright © 2001 by Project MIND, Inc.
All rights reserved, including the right to reproduce these flash cards or portions thereof in any form whatsoever.

The Mental Math Game

The students form two teams and come up to the bells two at a time. Upon looking at the math problem on a yellow card, they solve the problem mentally as fast as they can, usually within three seconds. The winner continues on while the loser moves to the end of his line. To be an intermediate champion, one must respond to three problems in a row correctly. After four intermediate champions are picked (depending on the size of your group, you must make sure that each student had a t least three chances), they are then entered into the second level of competitions with the green cards (more difficult problems.) To be a runner up for the grand champion title, competitors must also respond correctly to three problems in a row. Two runner-ups for the advanced level cards (red) are picked. They now compete for the title. The first person to respond to three red card problems in a row correctly is the grand champion.

Variations:

- ➢ The students compete in four areas: Mentally solve math problems with cards (visual aids), mentally solve math problems without cards, word problems, and equations (a string of problems to solve as the reader reads them.)
- ➢ The game can be played with your own class, another class, your grade level, or with other grade levels (fourth grade competing against fifth grade, third grade competing against fourth grade, etc.)
- ➢ If you have an advanced group, make sure that they use the cards for the next grade.
- ➢ Decimals and fractions can be added for third through fifth grade.

Pre-Kindergarten/Kindergarten:

- Level 1 – Yellow Cards: Number identification and shape identification
- Level 2 – Green Cards: Number identification (up to 100), identify the missing number, and adding and subtracting up to 5.
- Level 3 – Red Cards: number sequencing, and adding and subtracting up to 10
- Equations: strings of numbers which add and subtract up to 10
- Word problems: Simple one step, how many items? Adding or subtracting up to 10

First Grade:

- Level 1 – Yellow Cards: Adding and subtracting numbers up to 10
- Level 2 – Green Cards: Adding and subtracting two-digit numbers and adding three digit numbers
- Level 3 – Red Cards: Adding and subtracting three-digit numbers

Second Grade:

- Level 1 – Yellow Cards: Adding and subtracting two-digit numbers

- Level 2 – Green Cards: Adding and subtracting two-digit numbers with carrying and borrowing, and multiplication and division facts
- Level 3 – Red Cards: Adding and subtracting three-digit numbers with carrying and borrowing; two-digit multiplication

Third Grade:

- Level 1 – Yellow Cards: Adding and subtracting two-digit numbers with carrying and borrowing; single digit multiplication and division
- Level 2 – Green Cards: Adding and subtracting three-digit numbers with carrying and borrowing, and two-digit multiplication and division
- Level 3 – Red Cards: Adding and subtracting four-digit numbers with carrying and borrowing; three-digit multiplication and division

Fourth Grade:

- Level 1 – Yellow Cards: Adding, subtracting, multiplying, and dividing fourth grade level problem
- Level 2 – Green Cards: Adding, subtracting, multiplying, and dividing fourth grade level problems that are harder than Level 1
- Level 3 – Red Cards: Adding, subtracting, multiplying, and dividing multi-digit fifth grade level problems

Fifth Grade:

- Level 1 – Yellow Cards: Adding, subtracting, multiplying, and dividing fifth grade level problem
- Level 2 – Green Cards: Adding, subtracting, multiplying, and dividing fifth grade level problems that are harder than Level 1
- Level 3 – Red Cards: Adding and subtracting six digit numbers with carrying and borrowing, and multiplying and dividing multi-digit problems

$$\begin{array}{r} 82 \\ +\ 11 \\ \hline \end{array}$$

$$\begin{array}{r} 72 \\ +\ 69 \\ \hline \end{array}$$

The Mental Math Game

The students form two teams and come up to the bells two at a time. Upon looking at the math problem on a yellow card, they solve the problem mentally as fast as they can, usually within three seconds. The winner continues on while the loser moves to the end of his line. To be an intermediate champion, one must respond to three problems in a row correctly. After four intermediate champions are picked (depending on the size of your group, you must make sure that each student had a t least three chances), they are then entered into the second level of competitions with the green cards (more difficult problems.) To be a runner up for the grand champion title, competitors must also respond correctly to three problems in a row. Two runner-ups for the advanced level cards (red) are picked. They now compete for the title. The first person to respond to three red card problems in a row correctly is the grand champion.

Variations:

- ➢ The students compete in four areas: Mentally solve math problems with cards (visual aids), mentally solve math problems without cards, word problems, and equations (a string of problems to solve as the reader reads them.)
- ➢ The game can be played with your own class, another class, your grade level, or with other grade levels (fourth grade competing against fifth grade, third grade competing against fourth grade, etc.)
- ➢ If you have an advanced group, make sure that they use the cards for the next grade.
- ➢ Decimals and fractions can be added for third through fifth grade.

Pre-Kindergarten/Kindergarten:

- ▪ Level 1 – Yellow Cards: Number identification and shape identification
- ▪ Level 2 – Green Cards: Number identification (up to 100), identify the missing number, and adding and subtracting up to 5.
- ▪ Level 3 – Red Cards: number sequencing, and adding and subtracting up to 10
- ▪ Equations: strings of numbers which add and subtract up to 10
- ▪ Word problems: Simple one step, how many items? Adding or subtracting up to 10

First Grade:

- ▪ Level 1 – Yellow Cards: Adding and subtracting numbers up to 10
- ▪ Level 2 – Green Cards: Adding and subtracting two-digit numbers and adding three digit numbers
- ▪ Level 3 – Red Cards: Adding and subtracting three-digit numbers

Second Grade:

- ▪ Level 1 – Yellow Cards: Adding and subtracting two-digit numbers

- Level 2 – Green Cards: Adding and subtracting two-digit numbers with carrying and borrowing, and multiplication and division facts
- Level 3 – Red Cards: Adding and subtracting three-digit numbers with carrying and borrowing; two-digit multiplication

Third Grade:

- Level 1 – Yellow Cards: Adding and subtracting two-digit numbers with carrying and borrowing; single digit multiplication and division
- Level 2 – Green Cards: Adding and subtracting three-digit numbers with carrying and borrowing, and two-digit multiplication and division
- Level 3 – Red Cards: Adding and subtracting four-digit numbers with carrying and borrowing; three-digit multiplication and division

Fourth Grade:

- Level 1 – Yellow Cards: Adding, subtracting, multiplying, and dividing fourth grade level problem
- Level 2 – Green Cards: Adding, subtracting, multiplying, and dividing fourth grade level problems that are harder than Level 1
- Level 3 – Red Cards: Adding, subtracting, multiplying, and dividing multi-digit fifth grade level problems

Fifth Grade:

- Level 1 – Yellow Cards: Adding, subtracting, multiplying, and dividing fifth grade level problem
- Level 2 – Green Cards: Adding, subtracting, multiplying, and dividing fifth grade level problems that are harder than Level 1
- Level 3 – Red Cards: Adding and subtracting six digit numbers with carrying and borrowing, and multiplying and dividing multi-digit problems

```
  82
+ 11
────
  93
════
```

```
   72
+  69
┌─────┐
│ 141 │
└─────┘
```

```
  634        577
+ 178      + 346
_____      _____
```

$$\begin{array}{r} 634 \\ +\ 178 \\ \hline 812 \\ \hline \end{array}$$

Project MIND
Fourth - Yellow

$$\begin{array}{r} 577 \\ +\ 346 \\ \hline 923 \\ \hline \end{array}$$

Project MIND
Fourth - Yellow

$$\begin{array}{r} 623 \\ +\ 293 \\ \hline \end{array}$$

$$\begin{array}{r} 769 \\ +\ 901 \\ \hline \end{array}$$

```
  623
+ 293
─────
  916
```

```
   769
+  901
──────
  1670
```

$$\begin{array}{r} 3,529 \\ +\ 4,486 \\ \hline \end{array}$$

$$\begin{array}{r} 5,451 \\ +\ 5,498 \\ \hline \end{array}$$

```
        3,529
  +     4,486
        8,015
```

Project MIND
Fourth - Yellow

```
        5,451
  +     5,498
       10,949
```

Project MIND
Fourth - Yellow

$$
\begin{array}{r}
1,894 \\
+\ 1,087 \\
\hline
\end{array}
$$

$$
\begin{array}{r}
6,720 \\
+\ 2,956 \\
\hline
\end{array}
$$

$$\begin{array}{r} 1,894 \\ +1,087 \\ \hline 2,981 \end{array}$$

Project MIND
Fourth - Yellow

$$\begin{array}{r} 6,720 \\ +2,956 \\ \hline 9,676 \end{array}$$

Project MIND
Fourth - Yellow

$$
\begin{array}{r}
8{,}936 \\
+5{,}830 \\
\hline
\end{array}
$$

$$
\begin{array}{r}
9{,}356 \\
+3{,}782 \\
\hline
\end{array}
$$

$$\begin{array}{r} 8{,}936 \\ +5{,}830 \\ \hline 14{,}766 \\ \hline\hline \end{array}$$

Project MIND
Fourth - Yellow

$$\begin{array}{r} 9{,}356 \\ +3{,}782 \\ \hline 13{,}138 \\ \hline\hline \end{array}$$

Project MIND
Fourth - Yellow

$$95 - 67$$

$$88 - 39$$

```
  95
- 67
────
  28
```

```
  88
- 39
────
  49
```

$$473 - 92$$

$$334 - 19$$

```
  473
-  92
─────
  381
```

```
  34
- 19
────
  15
```

$$\begin{array}{r} 804 \\ -\ 75 \\ \hline \end{array}$$

$$\begin{array}{r} 511 \\ -\ 67 \\ \hline \end{array}$$

804
75

729

511
- 67

444

```
    907        332
  -  532     -  148
```

907
532
-
375

332
148
-
184

$$947 - 654$$

$$927 - 543$$

```
  947
- 654
─────
  293
```

```
  927
- 543
─────
  384
```

$$\begin{array}{r} 1,596 \\ -898 \\ \hline \end{array}$$

$$\begin{array}{r} 8,439 \\ -7,893 \\ \hline \end{array}$$

$$\begin{array}{r} 1{,}596 \\ -898 \\ \hline 698 \\ \hline\hline \end{array}$$

Project MIND
Fourth - Yellow

$$\begin{array}{r} 8{,}439 \\ -7{,}893 \\ \hline 546 \\ \hline\hline \end{array}$$

Project MIND
Fourth - Yellow

$$
\begin{array}{r}
5{,}007 \\
-3{,}846 \\
\hline
\end{array}
$$

$$
\begin{array}{r}
7{,}261 \\
-2{,}948 \\
\hline
\end{array}
$$

$$\begin{array}{r} 5{,}007 \\ -\ \underline{3{,}846} \\ \underline{\underline{1{,}161}} \end{array}$$

Project MIND
Fourth - Yellow

$$\begin{array}{r} 7{,}261 \\ -\ \underline{2{,}948} \\ \underline{\underline{4{,}313}} \end{array}$$

Project MIND
Fourth - Yellow

$$14 \times 5$$

$$19 \times 3$$

14
x 5
‒‒‒‒
70

19
x 3
‒‒‒‒
57

$$\begin{array}{r} 24 \\ \times\ 3 \\ \hline \end{array}$$

$$\begin{array}{r} 26 \\ \times\ 2 \\ \hline \end{array}$$

24
x 3

72

26
x 2

52

$$38 \times 2$$

$$32 \times 3$$

$$\begin{array}{r} 38 \\ \times\ 2 \\ \hline 76 \end{array}$$

$$\begin{array}{r} 32 \\ \times\ 3 \\ \hline 96 \end{array}$$

$$\begin{array}{r} 45 \\ \times\ 4 \\ \hline \end{array}$$

$$\begin{array}{r} 41 \\ \times\ 2 \\ \hline \end{array}$$

```
  45
x  4
―――
 180
```

```
     41
x     2
―――――
    82
```

44
× 6

46
× 8

```
  44
x  9
――――
 396
```

Project MIND
Fourth - Yellow

```
  46
x  8
――――
 368
```

Project MIND
Fourth - Yellow

$$\begin{array}{r} 98 \\ \times\ 9 \\ \hline \end{array}$$

$$\begin{array}{r} 48 \\ \times\ 7 \\ \hline \end{array}$$

```
    98
x    6
   588
```

```
    48
x    7
   336
```

$$91 \div 7 = $$

$$90 \div 3 = $$

$$\begin{array}{r} 91 \\ \div \quad 7 \\ \hline 13 \end{array}$$

Project MIND
Fourth - Yellow

$$\begin{array}{r} 90 \\ \div \quad 3 \\ \hline 30 \end{array}$$

Project MIND
Fourth - Yellow

$$84 \div 6$$

$$76 \div 4$$

$$\begin{array}{r} 84 \\ \div\ 6 \\ \hline 14 \\ \hline\hline \end{array}$$

Project MIND
Fourth - Yellow

$$\begin{array}{r} 76 \\ \div\ 4 \\ \hline 19 \\ \hline\hline \end{array}$$

Project MIND
Fourth - Yellow

$$78 \div 6$$

$$90 \div 5$$

$$\begin{array}{r} 78 \\ \div 6 \\ \hline 13 \end{array}$$

Project MIND
Fourth - Yellow

$$\begin{array}{r} 90 \\ \div 5 \\ \hline 18 \end{array}$$

Project MIND
Fourth - Yellow

$$96 \div 3$$

$$70 \div 7$$

$$\begin{array}{r} 96 \\ \div \quad 3 \\ \hline 32 \end{array}$$

Project MIND
Fourth - Yellow

$$\begin{array}{r} 70 \\ \div \quad 7 \\ \hline 10 \end{array}$$

Project MIND
Fourth - Yellow

$$70 \div 5$$

$$76 \div 2$$

$$\begin{array}{r} 70 \\ \div\ \ \ 5 \\ \hline 14 \end{array}$$

..

Project MIND
Fourth - Yellow

$$\begin{array}{r} 76 \\ \div\ \ \ 2 \\ \hline 38 \end{array}$$

$$108 \div 4$$

$$88 \div 2$$

$$\begin{array}{r} 108 \\ \div\ \quad 4 \\ \hline 27 \end{array}$$

Project MIND
Fourth - Yellow

$$\begin{array}{r} 88 \\ \div\ \quad 2 \\ \hline 44 \end{array}$$

Project MIND
Fourth - Yellow

$$907 + 532$$

$$907 + 129$$

```
   907
+  532
──────
  1439
```

```
   907
+  129
──────
  1036
```

$$698 + 403$$

$$387 + 96$$

$$\begin{array}{r} 698 \\ 403 \\ + \\ \hline 1101 \end{array}$$

$$\begin{array}{r} 387 \\ 96 \\ + \\ \hline 483 \end{array}$$

$$206 + 89$$

$$470 + 45$$

$$\begin{array}{r} 206 \\ +89 \\ \hline 295 \end{array}$$

$$\begin{array}{r} 470 \\ +45 \\ \hline \mathbf{515} \end{array}$$

$$26,004$$
$$+\ \ 14,832$$
___

$$7,043$$
$$+\ \ 2,462$$
___

$$\begin{array}{r} 26{,}004 \\ +\ 14{,}832 \\ \hline 40{,}836 \end{array}$$

Project MIND
Fourth - Green

$$\begin{array}{r} 7{,}043 \\ +\ 2{,}462 \\ \hline 9{,}505 \end{array}$$

Project MIND
Fourth - Green

$$
\begin{array}{r}
8,395 \\
+5,201 \\
\hline
\end{array}
$$

$$
\begin{array}{r}
6,543 \\
+5,678 \\
\hline
\end{array}
$$

$$\begin{array}{r} 8,395 \\ +\ \underline{5,201} \\ \underline{\underline{13,596}} \end{array}$$

Project MIND
Fourth - Green

$$\begin{array}{r} 6,543 \\ +\ \underline{5,678} \\ \underline{\underline{12,221}} \end{array}$$

Project MIND
Fourth - Green

$$\begin{array}{r} 3{,}172 \\ +\ 2{,}883 \\ \hline \end{array}$$

$$\begin{array}{r} 5{,}089 \\ +\ \ 489 \\ \hline \end{array}$$

$$\begin{array}{r} 3{,}172 \\ +2{,}883 \\ \hline 6{,}055 \end{array}$$

Project MIND
Fourth - Green

$$\begin{array}{r} 5{,}089 \\ +489 \\ \hline 5{,}578 \end{array}$$

Project MIND
Fourth - Green

$$
\begin{array}{r}
62{,}893 \\
+\quad 22{,}345 \\
\hline
\end{array}
$$

$$
\begin{array}{r}
45{,}298 \\
+\quad 29{,}396 \\
\hline
\end{array}
$$

$$\begin{array}{r} 62,893 \\ +\ \underline{22,345} \\ \underline{\underline{85,238}} \end{array}$$

Project MIND
Fourth - Green

$$\begin{array}{r} 45,298 \\ +\ \underline{29,396} \\ \underline{\underline{74,694}} \end{array}$$

Project MIND
Fourth - Green

$$300 - 146$$

$$42 - 19$$

```
  300
- 146
  154
```

```
   42
 - 19
   23
```

```
   698        600
-  403      - 478
------      ------
```

```
  698
-
  403
  295
```

Project MIND
Fourth - Green

```
  600
-
  478
  122
```

Project MIND
Fourth - Green

$$\begin{array}{r} 1,406 \\ -90 \\ \hline \end{array}$$

$$\begin{array}{r} 7,010 \\ -4,121 \\ \hline \end{array}$$

$$
\begin{array}{r}
1,406 \\
-90 \\
\hline
1,316 \\
\hline
\end{array}
$$

Project MIND
Fourth - Green

$$
\begin{array}{r}
7,010 \\
-4,121 \\
\hline
2,889 \\
\hline
\end{array}
$$

$$\begin{array}{r} 5{,}691 \\ -3{,}218 \\ \hline \end{array}$$

$$\begin{array}{r} 4{,}702 \\ -1{,}792 \\ \hline \end{array}$$

$$\begin{array}{r} 5{,}691 \\ -3{,}218 \\ \hline 2{,}473 \\ \hline \end{array}$$

Project MIND
Fourth - Green

$$\begin{array}{r} 4{,}702 \\ -1{,}792 \\ \hline 2{,}910 \\ \hline \end{array}$$

Project MIND
Fourth - Green

$$
\begin{array}{r}
9,357 \\
- \quad 7,473 \\
\hline
\end{array}
$$

$$
\begin{array}{r}
78,560 \\
- \quad 24,967 \\
\hline
\end{array}
$$

$$
\begin{array}{r}
9{,}357 \\
-\ \ 7{,}473 \\
\hline
1{,}884 \\
\hline
\end{array}
$$

Project MIND
Fourth - Green

$$
\begin{array}{r}
78{,}560 \\
-\ \ 24{,}967 \\
\hline
53{,}593 \\
\hline
\end{array}
$$

Project MIND
Fourth - Green

22
× 6

28
× 6

$$\begin{array}{r} 22 \\ \times\ \ 6 \\ \hline 132 \end{array}$$

$$\begin{array}{r} 28 \\ \times\ \ 9 \\ \hline \boxed{252} \end{array}$$

$$27 \times 6$$

$$29 \times 7$$

$$\begin{array}{r} 27 \\ \times\ \ 6 \\ \hline 162 \end{array}$$

$$\begin{array}{r} 29 \\ \times\ \ 7 \\ \hline \boxed{203} \end{array}$$

$$\begin{array}{r} 37 \\ \times\ 9 \\ \hline \end{array}$$

$$\begin{array}{r} 38 \\ \times\ 7 \\ \hline \end{array}$$

$$
\begin{array}{r}
37 \\
\times\ 9 \\
\hline
333
\end{array}
$$

$$
\begin{array}{r}
38 \\
\times\ 7 \\
\hline
266
\end{array}
$$

$$37 \times 5$$

$$38 \times 4$$

```
    37
  x  5
  _____
   185
```

```
    38
  x  4
  _____
   152
```

$$41 \times 5$$

$$39 \times 7$$

41
x 5

205

39
x 7

273

$$\begin{array}{r} 59 \\ \times\ 3 \\ \hline \end{array}$$

$$\begin{array}{r} 46 \\ \times\ 8 \\ \hline \end{array}$$

```
   59
 x  3
─────
  177
```

```
   46
 x  8
─────
 368
```

$$63 \times 7$$

$$54 \times 7$$

63
x 7
―――
441

54
x 7
―――
378

$$7 \times 79$$

$$8 \times 77$$

```
    79
  x  7
  -----
   553
```

```
    77
  x  8
  -----
   616
```

$$84 \times 4$$

$$72 \times 7$$

```
  84
x  4
―――
 336
```

```
  72
x  7
―――
 504
```

$$95 \times 8$$

$$94 \times 7$$

$$\begin{array}{r} 95 \\ \times\ 8 \\ \hline 760 \end{array}$$

$$\begin{array}{r} 94 \\ \times\ 7 \\ \hline 658 \end{array}$$

$$60 \div 6$$

$$250 \div 5$$

$$60 \div 6 = 10$$

Project MIND
Fourth - Green

$$250 \div 5 = 50$$

Project MIND
Fourth - Green

$$160 \div 5$$

$$469 \div 7$$

160 ÷ 5 = 32

Project MIND
Fourth - Green

469 ÷ 7 = 67

Project MIND
Fourth - Green

315 ÷ 5

688 ÷ 8

$$315 \div 5 = 63$$

Project MIND
Fourth - Green

$$688 \div 8 = 86$$

Project MIND
Fourth - Green

$$288 \div 4$$

$$594 \div 9$$

288 ÷ 4 = 72

Project MIND
Fourth - Green

594 ÷ 9 = 66

Project MIND
Fourth - Green

$$395 \div 5$$

$$768 \div 8$$

$$395 \div 5 = 79$$

Project MIND
Fourth - Green

$$768 \div 8 = 96$$

Project MIND
Fourth - Green

$$552 \div 6$$

$$469 \div 7$$

$$552 \div 6 = 92$$

Project MIND
Fourth - Green

$$469 \div 7 = 67$$

Project MIND
Fourth - Green

$$
\begin{array}{r}
3{,}784 \\
+\quad 1{,}590 \\
\hline
\end{array}
$$

$$
\begin{array}{r}
8{,}920 \\
+\quad 4{,}012 \\
\hline
\end{array}
$$

$$\begin{array}{r} 3{,}784 \\ +\ 1{,}590 \\ \hline 5{,}374 \end{array}$$

Project MIND
Fourth - Red

$$\begin{array}{r} 8{,}920 \\ +\ 4{,}012 \\ \hline 12{,}932 \end{array}$$

Project MIND
Fourth - Red

$$\begin{array}{r} 1,748 \\ +299 \\ \hline \end{array}$$

$$\begin{array}{r} 6,397 \\ +745 \\ \hline \end{array}$$

$$\begin{array}{r} 1{,}748 \\ +299 \\ \hline 2{,}047 \end{array}$$

Project MIND
Fourth - Red

$$\begin{array}{r} 6{,}397 \\ +745 \\ \hline 7{,}142 \end{array}$$

Project MIND
Fourth - Red

$$\begin{array}{r} 2,940 \\ +1,945 \\ \hline \end{array}$$

$$\begin{array}{r} 5,421 \\ +2,749 \\ \hline \end{array}$$

$$\begin{array}{r} 2{,}940 \\ +1{,}945 \\ \hline 4{,}885 \\ \hline\hline \end{array}$$

Project MIND
Fourth - Red

$$\begin{array}{r} 5{,}421 \\ +2{,}749 \\ \hline 8{,}170 \\ \hline\hline \end{array}$$

Project MIND
Fourth - Red

$$
\begin{array}{r}
56{,}902 \\
+\quad 1{,}693 \\
\hline
\end{array}
$$

$$
\begin{array}{r}
76{,}208 \\
+\quad 1{,}984 \\
\hline
\end{array}
$$

$$\begin{array}{r} 56,902 \\ +1,693 \\ \hline 58,595 \end{array}$$

Project MIND
Fourth - Red

$$\begin{array}{r} 76,208 \\ +1,984 \\ \hline 78,192 \end{array}$$

Project MIND
Fourth - Red

$$\begin{array}{r} 61,923 \\ +9,744 \\ \hline \end{array}$$

$$\begin{array}{r} 54,866 \\ +4,950 \\ \hline \end{array}$$

$$\begin{array}{r} 61,923 \\ +9,744 \\ \hline 71,667 \\ \hline \end{array}$$

Project MIND
Fourth - Red

$$\begin{array}{r} 54,866 \\ +4,950 \\ \hline 59,816 \\ \hline \end{array}$$

Project MIND
Fourth - Red

$$
\begin{array}{r}
39,211 \\
+ \quad 12,799 \\
\hline
\end{array}
$$

$$
\begin{array}{r}
94,875 \\
+ \quad 20,655 \\
\hline
\end{array}
$$

$$\begin{array}{r} 39{,}211 \\ +\ \ 12{,}799 \\ \hline 52{,}010 \\ \hline \end{array}$$

Project MIND
Fourth - Red

$$\begin{array}{r} 94{,}875 \\ +\ \ 20{,}655 \\ \hline 115{,}530 \\ \hline \end{array}$$

Project MIND
Fourth - Red

$$74,863$$
$$-199$$
_____

$$43,902$$
$$-748$$
_____

$$74,863$$
$$-\ \ \ \ \ 199$$
$$74,664$$

Project MIND
Fourth - Red

$$43,902$$
$$-\ \ \ \ \ 748$$
$$43,154$$

Project MIND
Fourth - Red

$$
\begin{array}{r}
58{,}271 \\
-\quad 1{,}589 \\
\hline
\end{array}
$$

$$
\begin{array}{r}
69{,}102 \\
-\quad 8{,}476 \\
\hline
\end{array}
$$

$$\begin{array}{r} 58,271 \\ - \quad 1,589 \\ \hline 56,682 \end{array}$$

Project MIND
Fourth - Red

$$\begin{array}{r} 69,102 \\ - \quad 8,476 \\ \hline 60,626 \end{array}$$

Project MIND
Fourth - Red

$$\begin{array}{r} 85{,}320 \\ -9{,}789 \\ \hline \end{array}$$

$$\begin{array}{r} 95{,}193 \\ -7{,}619 \\ \hline \end{array}$$

$$\begin{array}{r} 85,320 \\ -9,789 \\ \hline 75,531 \\ \hline\hline \end{array}$$

Project MIND
Fourth - Red

$$\begin{array}{r} 95,193 \\ -7,619 \\ \hline 87,574 \\ \hline\hline \end{array}$$

Project MIND
Fourth - Red

$$
\begin{array}{r}
75{,}021 \\
-\quad 8{,}124 \\
\hline
\end{array}
$$

$$
\begin{array}{r}
61{,}096 \\
-\quad 3{,}659 \\
\hline
\end{array}
$$

```
      75,021
-      8,124
      66,897
```

Project MIND
Fourth - Red

```
      61,096
-      3,659
      57,437
```

Project MIND
Fourth - Red

$$24{,}798$$
$$-\quad 7{,}050$$
_____

$$81{,}560$$
$$-\quad 9{,}799$$
_____

$$
\begin{array}{r}
24{,}798 \\
-\ \underline{7{,}050} \\
\underline{17{,}748}
\end{array}
$$

Project MIND
Fourth - Red

$$
\begin{array}{r}
81{,}560 \\
-\ \underline{9{,}799} \\
\underline{71{,}761}
\end{array}
$$

Project MIND
Fourth - Red

$$419 \times 3$$

$$124 \times 6$$

```
  419
x   3
─────
 1257
```

```
  124
x   6
─────
  744
```

266
x 3

143
x 7

266
x   3
———
798

143
x   7
———
1001

$$937 \times 2$$

$$926 \times 7$$

$$\begin{array}{r} 937 \\ \times \quad 2 \\ \hline 1874 \end{array}$$

Project MIND
Fourth - Red

$$\begin{array}{r} 926 \\ \times \quad 7 \\ \hline 6482 \end{array}$$

Project MIND
Fourth - Red

$$305 \times 3$$

$$218 \times 8$$

```
  305
x   3
─────
  915
```

```
  218
x    8
──────
 1744
```

$$278 \times 5$$

$$208 \times 3$$

$$\begin{array}{r} 278 \\ \times \quad 5 \\ \hline 1390 \end{array}$$

$$\begin{array}{r} 208 \\ \times \quad 3 \\ \hline 624 \end{array}$$

$$
\begin{array}{r}
129 \\
\times\ 4 \\
\hline
\end{array}
$$

$$
\begin{array}{r}
714 \\
\times\ 6 \\
\hline
\end{array}
$$

$$\begin{array}{r} 129 \\ \times\ \ 4 \\ \hline 516 \end{array}$$

$$\begin{array}{r} 714 \\ \times\ \ 6 \\ \hline 4284 \end{array}$$

$$315 \times 4$$

$$393 \times 5$$

```
  315
x   4
─────
 1260
═════
```

```
  393
x   5
─────
 1965
═════
```

$$\begin{array}{r} 329 \\ \times 3 \\ \hline \end{array}$$

$$\begin{array}{r} 907 \\ \times 8 \\ \hline \end{array}$$

```
  329
x   3
─────
  987
═════
```

```
  907
x   8
──────
 7256
══════
```

$$419 \times 2$$

$$178 \times 5$$

```
  419
x   2
─────
  838
═════
```

```
  178
x   5
─────
  890
═════
```

$$3,052 \div 7$$

$$2,552 \div 9$$

$$3,052 \div 7 = 436$$

Project MIND
Fourth - Red

$$2,552 \div 9 = 283.56$$

Project MIND
Fourth - Red

$$1,812 \div 4$$

$$951 \div 3$$

$$1{,}812 \div 4 = 453$$

Project MIND
Fourth - Red

$$951 \div 3 = 317$$

Project MIND
Fourth - Red

1,120 ÷ 5

5,216 ÷ 8

$$1{,}120 \div 5 = 224$$

Project MIND
Fourth - Red

$$5{,}216 \div 8 = 652$$

Project MIND
Fourth - Red

1,036 ÷ 7

735 ÷ 5

$$1,036 \div 7 = 148$$

Project MIND
Fourth - Red

$$735 \div 5 = 147$$

Project MIND
Fourth - Red

1,416 ÷ 3

7,425 ÷ 9

$$1{,}416 \div 3 = 472$$

Project MIND
Fourth - Red

$$7{,}425 \div 9 = 825$$

Project MIND
Fourth - Red

$$2{,}456 \div 4$$

$$2{,}760 \div 5$$

$$2{,}456 \div 4 = 614$$

Project MIND
Fourth - Red

$$2{,}760 \div 5 = 552$$

Project MIND
Fourth - Red

$$1,125 \div 3$$

$$1,092 \div 6$$

$$1{,}125 \div 3 = 375$$

Project MIND
Fourth - Red

$$1{,}092 \div 6 = 182$$

Project MIND
Fourth - Red

3,258 ÷ 9

3,052 ÷ 7

$$3{,}258 \div 9 = 362$$

Project MIND
Fourth - Red

$$3{,}052 \div 7 = 436$$

Project MIND
Fourth - Red